Waffle Science

-by Bill Watkins

Traveling Poet Press – Livingston, Montana 2017

2

Part One:
Religio-Spiritual Science

4

What's the difference between *Spiritual* and *Religious*? I learned once at a place people visit to quit drinking alcohol that one could be spiritual and not be religious, and vice versa.

Alcohol sold as a drink has some scientific problems, if not religious ones.

C_2H_5OH is the chemical formula for ethyl alcohol, a substance that results from "fermentation"—and that some choose to drink.

It hurts me to think about it as I lost friends to alcohol-related accidents and depression.

Depression can result from the addiction some feel to drinking alcohol, C_2H_5OH – a substance dictionaries define as a "colorless, volatile, flammable liquid known as toxic."

One thing I have noticed on the subject is that a majority of people who drink this flammable, toxic liquid have never researched what alcohol actually is.

I think it is unwise and scientifically unsound to put something in our bodies we know nothing about.

Religiously, if one were to look at say, the Ten Commandments, I think it is true that some drink alcohol or use drugs instead of dedicating themselves to one God.

Both science and religion seem against alcohol and drug use, but some people will probably always use them.

Society might be wise, though, to treat

all alcohol and drugs like cigarettes: *Beware* of them, have no advertising for them on television.

In fact, Cigarette TV Ads were banned way back in 1970 by President Richard Nixon, who cited health reasons.

Why should alcohol and drug ads be allowed today?

Do you think it is right that children or

addicted people should ever see these ads on their television sets?

Waffle does not.

Part Two:
The Speed of Life

12

Light travels at 186,000 miles per second, which means that a beam of light could travel around the world about seven and a half times *IN ONE SINGLE SECOND*.

Some think there is evidence that God does not exist; that there is no Higher Power that runs things; that we are alone, and Science can explain all things.

Science seems to prove the existence of a Power greater than ourselves,

Waffle proposes, as he can list
so many things over which humans
have no control.

We did not make the earth, the rivers,
the valleys and the mountains.

We did not make all the animals,
the sky, the air.

Life emerged out of one big
bang, some scientists think. But
that does not explain what was there
before the bang.

Waffle likes "Big *Yang*" to explain the beginning of the Universe, using the Tao Te Ching's theory of opposites, Yin and Yang.

There is no matter without the absence of matter.

Likewise, there is no void or vacuum without matter to define it.

There is no Big Bang without a Big Dud; and no Big Dud without a great Big Bang.

There is an equal and opposite force to all things, said Isaac Newton—which makes sense, and in the end proves Atheism is okay after all.

Part Three:
Pre-School Algebra and Infinity

Two plus Two does not always equal four.

It depends what you are adding.

Likewise, two plus two can equal to twenty-two if you are adding 5.5's.

2(5.5) + 2(5.5) = 22.

Challenging everything is a good mood for learning and philosophy.

When one is ready to make a decision

or a statement, write it down to leave philosophy and enter into Art, Science, Creative Writing, or Math.

1=0=infinity is an old equation Waffle came up with one day. So much in an equation depends on how we define things with words.

In that sense, math and science are very poetic, the best poems so
truthful as to be scientific.

One is a building block for all things, "All" or Infinity.

Zero is something, or one thing… or "one."

$E=mc^2$ is incomplete to me without

a "one" in there to represent zero and infinity, therefore E=mc²+1 is a better equation for a poet like me.

****

Big Yang

The wolf sees what scientists don't,
the full circle, Yin and Yang more than
Big Bang, this is the beginning, the end:

The same thing.

1, 2, 3 the dream of all who see, two
plus two? Depends what you're adding.
The cloth and the sun eclipsed by
spider webbing truth!

One thing remains after the dust settles:

Infinity.

Or does it all come to zero, back full circle
to "I don't know." The wolf howls angry
messages of love at the moon—the reverse of
sun, the inverse of none:

We are born...

Birthday rhymes with Earth day as we see out
the shadow call it soul to shade, *los musicos* preparing
to blast to the bull's demise but not before a final
surprise, the Earth cringing, its name a lullaby...

Sleep

Chess

Don't do, think, step back,
go for the best, for the enemy of the best
is a pretty good move.

Then see your error, your best move failed
this can't be fair.

God's weaving a masterpiece, a poem
for all ages,

The Devil beating a steady beat,
calling men and women to destroy
and be destroyed.

Confusion, the Devil's game, rampant
in and out of doors as God provides
the steady, disciplined path of humble
supplication and joyful receipt.

The Devil picks holes in easy games and
lines of communication, convinces some
they don't need Higher Power just themselves.

A "lower power" beckons strongly from those
drumbeats, we are lambs to its slaughter

Unless…

Unless we turn around, notice every day the
kit of tools given to us to relate with the One
and his or her divine plan, that poem
magnificently weaved and being weaved.

Check this out before checkmated, dive off cliffs, rev up engines, speed down highways listening to the beat.

You feel high beating the system until it all collapses 'round the telephone pole, fire ablaze, bombs exploding you forgot to pray.

The best move: utter submission…

Then, then come strong knowing what's in charge, the game reverses, we've got a chance not to win but at something greater:

Peace of mind that we played it the right way.

We can sleep on that soundness. That's a win to Devil's chagrin we beat the beat and'll have to probably do it again

Good Luck with your moves.

Love...

—Coach Waffle

30

www.ingramcontent.com/pod-product-compliance
Lightning Source LLC
Chambersburg PA
CBHW041119180526
45172CB00001B/327